Beekeeping Essentials

Tools, Techniques, and Hive Protection

Sadiq Ali Al Qatari

Index

Chapter 3: Essential Beekeeping Tools in Action

3.1 Inspection Routines

- Proper Handling: Using the hive tool, smoker, and bee brush effectively.
- Common Inspections: Checking for brood health, honey production, and signs of pests.

3.2 Seasonal Tool Usage

- Spring and Summer: Frames, hive tool, smoker, and queen excluder.
- Fall and Winter: Hive insulation, varroa mite treatments, and mouse guards.

Chapter 4: Protecting Bee Health and Hive Safety

4.1 Natural Hive Threats

- Pests: Wax moths, varroa mites, and small hive beetles.
- Predators: Raccoons, skunks, and bears – strategies to prevent access to hives.

4.2 Environmental Threats

- Temperature Extremes: Insulating the hive and providing ventilation.
- Flood and Rain Protection: Hive stand placement and rain covers.

4.3 Chemical-Free Hive Health Strategies

- Integrated Pest Management (IPM): Reducing pests without chemicals.
- Beekeeping Hygiene: Regular cleaning, tool disinfection, and natural mite treatments.

Chapter 5: Hive Monitoring and Technology Tools

5.1 Digital Hive Scales

- Monitoring Weight: Real-time tracking of nectar and honey production.
- Seasonal Analysis: Analyzing colony productivity trends.

5.2 Temperature and Humidity Sensors

- Internal Conditions: Monitoring hive conditions remotely.
- Health Indicators: Using data to detect colony distress.

5.3 Infrared Imaging

- Thermal Maps: Identifying weak colonies and brood health without opening the hive.
- Detecting Robbing Behavior: Recognizing hive invaders through heat signatures.

Chapter 6: Hive Maintenance and Re-Queening

6.1 Frame and Foundation Management

- Replacing Worn Frames: Extending hive life by regular frame replacement.
- Wax Foundation: Installing fresh foundation for comb building.

6.2 Re-Queening

- Identifying a Failing Queen: Recognizing signs of queen decline.
- Successful Re-Queening: Techniques for introducing a new queen.

6.3 End-of-Season Hive Maintenance

- Harvesting Honey: Proper extraction and storing methods.
- Preparing for Winter: Insulating, providing food reserves, and securing the hive.

Chapter 7: Hive Security and Protecting Against Hive Theft

7.1 Preventing Hive Theft

- Location Tips: Placing hives out of sight and setting up barriers.
- Identification: Marking hives with unique IDs and installing tracking devices.

7.2 Legal and Insurance Options

- Bee Hive Insurance: Protecting your investment.
- Tracking Technology: Using GPS tracking to recover stolen hives.

Chapter 8: Sustainable Beekeeping and Community Impact

8.1 Eco-Friendly Beekeeping Practices

- Minimizing Chemicals: Organic pest control methods.
- Bee Health: Avoiding stress through sustainable practices.

8.2 Community Engagement and Pollinator Education

- Educating on Bees: Community outreach on the importance of bees.
- Hosting Workshops: Sharing knowledge on beekeeping basics and bee protection.

Chapter 1: The Essentials of Beekeeping Tools

Beekeeping is an ancient practice that requires not only patience and passion but also the right set of tools to ensure the safety of both the beekeeper and the bees. This chapter delves into the foundational tools that every beekeeper, whether a beginner or seasoned professional, should have in their arsenal. By understanding each tool's purpose, function, and the best practices for their use, beekeepers can effectively manage their hives, optimize honey production, and safeguard the well-being of their colonies.

1.1 Bee Suit and Protective Gear

The bee suit is arguably the most iconic piece of equipment in beekeeping. Designed for safety, comfort, and durability, the suit is the first line of defense against bee stings and other hive-related hazards. Beekeeping involves close interaction with large colonies of bees, and while bees are generally gentle, they will defend their home when they feel threatened.

Bee Suit

A full-coverage bee suit provides protection from stings and minimizes discomfort during hive inspections. A typical bee suit covers the entire body and is crafted from a breathable yet durable material, allowing airflow while also preventing bees from penetrating the fabric. Suits come in various materials, with the most popular being cotton and ventilated mesh fabric. Cotton suits are durable and easier to maintain, while ventilated suits are more comfortable for warm climates as they allow air circulation.

Another important feature is the veil. A veil is a head covering that protects the face and neck, the most sensitive areas susceptible to stings. The veil is usually attached to the suit or helmet and features a mesh design that allows the beekeeper to see clearly while keeping bees away from their face. The hooded veil style has become particularly popular because it prevents bees from entering through the neck area.

Gloves and Boots

When it comes to handling beekeeping equipment, gloves are indispensable. Beekeeping gloves are typically made from leather or thick fabric and offer a balance between dexterity and protection. Leather gloves are highly durable and prevent stings from penetrating the material, though they may lack flexibility. Fabric gloves, on the other hand, allow for more precise handling of hive tools but provide less sting protection.

Alongside gloves, beekeeping boots provide foot protection while working in the apiary. Bees can be unpredictable, and stepping on them accidentally can lead to defensive reactions. Boots should be sturdy, ideally made from leather or rubber, and fit snugly to prevent bees from crawling inside.

Benefits of Protective Gear

Protective gear not only prevents physical harm but also reduces stress while handling bees, allowing beekeepers to stay calm and focused. The added confidence that comes with wearing a suit, gloves, and boots creates a safer experience and allows beekeepers to interact with their colonies more effectively. Additionally, fewer defensive actions from the bees lead to a more harmonious hive environment, which is beneficial for hive health and productivity.

1.2 Hive Tool

The hive tool is an understated yet essential part of the beekeeper's toolkit. This multi-purpose tool is specifically designed to handle several critical functions during hive inspections, and its compact, durable design makes it indispensable.

Features

Typically made of metal, the hive tool has a flat, sharp end and a hook or curved end, allowing beekeepers to pry open hive boxes, separate frames, scrape off propolis, and perform various other tasks. It resembles a small crowbar, with a narrow blade that can easily slide into crevices within the hive structure.

Functions

Separating Hive Boxes: Beehives are made up of different stacked boxes, and bees use propolis (a resinous substance) to seal these boxes together. The hive tool helps break the propolis seal and allows beekeepers to lift and separate boxes easily.

Lifting Frames: During inspections, it's often necessary to lift frames to check on the brood, honey stores, and the general health of the colony. The hive tool allows beekeepers to carefully lift frames without disrupting the bees too much.

Scraping Propolis and Wax: Propolis can accumulate in inconvenient places, making it difficult to work with the hive. The hive tool's scraping edge helps remove excess propolis, keeping the hive clean and functional.

The hive tool's versatility makes it a central component of any beekeeper's kit, allowing them to perform a range of maintenance tasks efficiently and effectively.

1.2 Smoker

The smoker is perhaps the most critical tool for any beekeeper, as it plays a key role in managing bee behavior during hive inspections. By emitting controlled puffs of smoke, the smoker helps calm bees, reducing their defensiveness and allowing beekeepers to work safely around the hive.

Purpose

The smoker works by masking the bees' alarm pheromones. Bees communicate through these pheromones, and when they feel threatened, they release an alarm signal that triggers a defensive response from the rest of the colony. By dispersing smoke, beekeepers can disrupt this communication, keeping the bees calm and making it easier to inspect the hive.

Operation

A basic smoker consists of a metal canister that holds a burning fuel source, a lid with a spout to direct the smoke, and bellows that control the airflow. The bellows pump air into the smoker, fueling the combustion process and producing smoke. Beekeepers can adjust the amount and intensity of smoke by controlling the airflow, using just enough to calm the bees without overwhelming them.

Fuel Sources

Common smoker fuels include natural materials such as dried grass, pine needles, and burlap. Each fuel source has its advantages, with burlap being a popular choice due to its ability to produce cool, steady smoke that lasts longer. Natural materials also help maintain a mild smoke, which is gentle on the bees.

Best Practices

Using the smoker requires a careful approach. Beekeepers typically start by puffing smoke at the entrance of the hive, giving bees a warning and calming them before opening the hive. It's important to use just a few puffs of smoke – over-smoking can stress the colony and harm hive health. After the entrance, beekeepers may lightly smoke the tops of the frames before lifting them out. This helps keep bees calm throughout the inspection, allowing for smoother and safer hive management.

1.4 Bee Brush

The bee brush is a tool designed to gently remove bees from frames and other surfaces without harming or agitating them. Its soft bristles allow for a careful approach, giving beekeepers an alternative to more intrusive methods when moving bees out of the way.

Soft Bristles

The bee brush is typically made from soft synthetic or natural bristles that can gently sweep bees away without damaging their wings or exoskeleton. The softness of the bristles is critical, as harsher materials can harm bees and increase their stress levels.

Usage Tips

Using a bee brush requires a gentle, slow approach. Sudden movements or aggressive brushing can distress bees and cause defensive behavior. When brushing bees off a frame, it's best to use slow, sweeping motions that guide them off without direct contact, minimizing disturbance and promoting a calm hive environment.

1.5 Other Essential Tools

Beyond the primary tools, several additional tools can simplify beekeeping tasks and enhance hive management. These tools may not be used in every inspection but are highly valuable for specific maintenance activities.

Frame Gripper

A frame gripper is a device used to lift frames without directly handling them, providing an easier grip and minimizing the risk of crushing bees or damaging the comb. The tool's clamping mechanism securely holds the frame, allowing beekeepers to lift it smoothly for inspection. It's particularly useful when working with heavy frames full of honey or pollen.

Uncapping Knife

An uncapping knife is used during honey harvesting to slice open the wax caps that bees seal over honey cells. These knives are designed with serrated or heated blades to easily cut through the wax without damaging the comb beneath. In addition to traditional knives, electric uncapping knives are also available, offering a heated blade that glides through the wax with ease.

Queen Catcher and Marking Tools

These tools are helpful when it's necessary to mark or relocate the queen. A queen catcher is a small, hand-held cage that safely captures the queen without harming her. Marking pens, often non-toxic paint pens, are used to identify the queen by marking a small dot on her thorax, allowing beekeepers to easily locate her during inspections.

Benefits of Using Additional Tools

Additional tools can make hive maintenance more efficient and precise, allowing beekeepers to carry out specific tasks with less stress on the bees and themselves. By using tools designed for particular purposes, beekeepers minimize disruption within the hive, leading to a healthier and more productive colony.

Chapter 2: Understanding the Bee Hive Structure

A deep understanding of bee hive structure is fundamental for any beekeeper striving to maintain a productive and healthy colony. Each part of the hive plays a specific role in facilitating the bees' natural behaviors, promoting hive health, and ensuring honey production. In this chapter, we'll cover the core components of a standard bee hive, explore methods for protecting the hive's structure, and offer guidance on hive placement to create the most suitable environment for bees to thrive.

2.1 Parts of the Hive

A bee hive's structure is both efficient and intricate, designed to support the bees' daily needs, house the queen and brood, and store honey and pollen. The most commonly used hive design in modern beekeeping is the Langstroth hive, celebrated for its modular design and ease of use.

Hive Boxes and Frames

In a typical Langstroth hive, there are two main types of boxes used to organize and control the hive's internal structure: brood boxes and honey supers. These boxes are stacked vertically, allowing the bees to expand upward as the colony grows.

- **Brood Boxes:** These are the foundational boxes placed at the bottom of the hive where the queen lays eggs, creating the "brood nest." The brood boxes provide space for larvae to develop and are essential for maintaining a healthy bee population. Brood boxes are typically left undisturbed during honey harvests to prevent disrupting the brood.

- **Honey Supers:** Honey supers are shorter boxes added above the brood boxes. These boxes are where bees store their honey. Supers are specifically designed for easy removal during harvest time, allowing beekeepers to access honey stores without disturbing the brood. Honey supers can be added or removed as needed throughout the season, depending on the colony's productivity.

Within each box, frames are used to hold comb. Frames consist of a wooden or plastic framework on which bees build their wax comb. Comb cells serve as storage for honey, pollen, and brood, providing the necessary infrastructure for bee activities. Frames can be removed individually, allowing beekeepers to inspect and manage the hive more easily. This modularity, a defining feature of the Langstroth hive, is essential for routine inspections, pest management, and honey extraction.

Queen Excluder

The queen excluder is a flat, grid-like piece that is placed between the brood box and honey supers. Its primary purpose is to prevent the queen from accessing honey supers. The openings in the excluder are designed so that worker bees, which are smaller, can pass through, but the larger queen cannot. This separation ensures that brood-rearing is confined to the brood box and that honey supers remain free of brood.

Using a queen excluder simplifies honey harvesting by keeping the honey supers dedicated solely to honey storage, minimizing disruption of the brood area during inspections and extraction.

2.2 Protecting the Hive's Structure

Beekeepers are responsible for ensuring that hive structures remain durable and resilient. Exposure to the elements, pest activity, and natural wear can degrade hives over time. By taking proactive steps, beekeepers can maintain the longevity of their hive equipment and provide a stable, safe home for the colony.

Wood Conditioners

Since most Langstroth hives are made from wood, they require regular treatment to withstand rain, wind, and fluctuating temperatures. Wood conditioners and weatherproofing treatments can enhance the durability of hive boxes, extending their lifespan.

- **Wood Stains and Sealants:** Applying a non-toxic, bee-safe wood sealant can protect hive boxes from moisture, preventing rotting and warping. Bees are highly sensitive to chemical residues, so it's essential to use only natural or organic products that won't harm the colony.

- **Painting:** Painting the exterior of hive boxes with a non-toxic, weather-resistant paint can also add a layer of protection against moisture. White or light-colored paint helps reflect sunlight, keeping the hive cooler during warm weather. However, it's important to avoid painting the interior, as this could release fumes that may disturb the bees.

Seasonal Hive Inspections

Conducting thorough inspections each season is essential to maintaining the hive's structure. Different seasons present unique challenges that can impact hive integrity.

- **Spring Inspections:** After winter, it's crucial to inspect the hive for any signs of moisture damage or structural issues that might have occurred during cold weather. Frames and boxes should be checked for mold or rot, and any damaged parts should be replaced promptly.

- **Summer and Fall Inspections:** Warmer seasons increase the risk of pest infestations, especially wax moths and small hive beetles, which can damage the hive structure and disrupt the colony. During these inspections, beekeepers should check for pest activity, repair damaged comb, and ensure that the hive structure remains sound.

Consistent maintenance prevents small issues from escalating and helps sustain a stable, secure environment within the hive.

2.3 Optimizing Hive Placement

Choosing the right location for a bee hive is a foundational aspect of beekeeping. Hive placement can significantly impact the colony's productivity, health, and ability to withstand environmental stressors. Several key factors play into optimizing hive location, from shade and sunlight exposure to protection from the elements.

Shade and Sunlight

Bees thrive in locations that receive an ideal balance of sunlight and shade. Hive temperature is a critical factor in colony health, as it influences bee activity, brood development, and honey production.

- **Early Morning Sunlight:** Placing hives in an area that receives direct sunlight in the early morning can help stimulate bee activity. As the hive warms up, bees become active earlier in the day, allowing them to maximize foraging hours.

- **Partial Shade:** In particularly hot climates, hives should be positioned to receive some afternoon shade, which can prevent overheating. If natural shade isn't available, beekeepers can set up shade structures or place the hive in a location where trees provide partial cover. Overheating can lead to reduced productivity and increased stress on the bees, so careful planning around sunlight exposure is essential.

Weather Protection

Protecting the hive from extreme weather conditions ensures that bees can maintain a stable internal environment, which is crucial for the colony's overall well-being.

- **Wind Protection:** High winds can disrupt foraging activities and make it difficult for bees to access their hives. When placing hives, beekeepers should seek sheltered areas that provide wind protection, such as along natural windbreaks like tree lines, fences, or shrubs.

- **Rain and Snow:** In regions with heavy rainfall or snow, hives should be elevated on stands to prevent water from pooling around the base. Hive covers or slanted roofs are also recommended to allow water to run off and prevent leaks. Snow accumulation can block entrances, so in winter climates, beekeepers often create windbreaks to minimize snow buildup around the hive entrance.

- **Temperature Extremes:** Extreme temperatures, both hot and cold, can be detrimental to hive health. In very cold climates, additional insulation may be necessary to prevent the colony from freezing, while in extremely hot climates, hives may need extra ventilation to prevent overheating.

Additional Placement Considerations

In addition to sunlight, shade, and weather protection, beekeepers should consider factors such as accessibility, safety, and the surrounding environment.

- **Accessibility:** Hives should be easily accessible for regular inspections and honey harvesting, allowing beekeepers to move around the hive comfortably. Accessibility also plays a role in emergency situations, as quick access to hives can be crucial when dealing with potential issues like swarming or pest infestations.

- **Safety:** For the safety of the bees and humans, hives should be placed away from high-traffic areas. If placing hives near homes or public spaces, it's best to orient hive entrances away from pathways to minimize encounters with people.

- **Surrounding Flora:** Bees forage within a few miles of their hive, so placing hives near abundant and diverse floral sources ensures a steady supply of nectar and pollen. Natural areas with flowering plants, trees, and water sources create an ideal setting for honey production and colony growth.

By understanding hive structure and optimizing its placement, beekeepers set the foundation for a successful and sustainable apiary. Protecting the structural integrity of hives and choosing appropriate locations not only boosts productivity but also contributes to the well-being and longevity of the bee colonies.

Chapter 3: Essential Beekeeping Tools in Action

Mastering the tools of beekeeping goes beyond simply owning them; it involves understanding their purpose and developing skillful techniques for handling them effectively throughout each season. This chapter dives into key inspection routines and highlights seasonal tool usage to help beekeepers maintain a healthy, productive hive year-round. We'll discuss the proper handling of foundational tools during inspections and detail how seasonal demands affect the use of certain tools and treatments to support hive health through changing conditions.

3.1 Inspection Routines

Regular hive inspections are essential for assessing colony health, monitoring brood production, and identifying any signs of disease or pest infestation. During these inspections, beekeepers rely on essential tools, such as the hive tool, smoker, and bee brush, to interact with the hive in a way that minimizes stress on the bees. Inspections should be conducted with care and precision to ensure that the hive environment remains as undisturbed as possible.

- **Proper Handling:** Using the Hive Tool, Smoker, and Bee Brush Effectively

- **Hive Tool:** As the primary tool for opening and managing hives, the hive tool is indispensable during inspections. It has two main functions—prying and scraping:

- **Prying Open Hive Boxes:** Bees use propolis, a sticky resin, to seal cracks in the hive and secure hive boxes together. The hive tool's flat, wedged edge is perfect for gently separating hive boxes, allowing beekeepers to lift and inspect individual boxes without disrupting the comb.

- **Lifting Frames:** The hive tool can be used to loosen frames stuck to the hive walls, making them easier to lift and examine. When lifting frames, beekeepers should move slowly and steadily to avoid jarring the bees.

- **Scraping Propolis and Wax:** Bees produce excess propolis and wax, which can obstruct frames and other hive components. Scraping excess material keeps the hive interior tidy and prevents clogging.

- **Smoker:** The smoker is essential for calming bees during inspections by masking alarm pheromones and creating a sense of calm within the hive.

- **Igniting and Using Fuel:** To produce smoke, beekeepers typically use natural fuels like pine needles, wood shavings, or dried grass. The smoker should be ignited and kept ready at the hive entrance, allowing smoke to fill the hive gradually.

- **Applying Smoke:** Applying small puffs of smoke at the entrance and near the top of the hive helps keep the bee's calm. Excessive smoke can cause stress, so it's best to use short bursts sparingly, pausing to gauge the bees' response.

- **Bee Brush:** The bee brush is used to gently move bees away from frames or hive components. Since bees are sensitive to sudden movements, the brush must be used with a light hand to avoid agitating them.

- **Brushing Technique:** Beekeepers should use smooth, light strokes, brushing bees away from the frames or boxes they need to access. Brushing too forcefully can crush or injure bees, causing unnecessary distress.

Common Inspections: Checking for Brood Health, Honey Production, and Signs of Pests

Each inspection provides valuable insights into the colony's well-being, from brood development to honey stores and pest management. Some key areas of focus during routine checks include:

- **Brood Health:** Inspecting brood frames helps beekeepers monitor the colony's reproductive health. Healthy brood patterns—solid, consistent clusters of capped brood cells—indicate a well-laying queen and a stable population. Spotting empty cells, scattered brood patterns, or dead larvae may signal issues such as disease, pest interference, or a failing queen.

- **Honey Production:** Checking honey frames provides insight into the hive's food stores. During nectar flows, beekeepers should see an increase in honey production, with honey frames filling and becoming capped. Monitoring honey stores is especially important leading into winter, as adequate honey reserves are crucial for the colony's survival through colder months.

- **Signs of Pests:** Identifying pests early is key to maintaining hive health. Common pests, such as varroa mites, small hive beetles, and wax moths, can cause significant damage if left untreated. During inspections, beekeepers should carefully examine frames, box interiors, and the bottom board for signs of pest activity, such as mite infestations on bees or larvae, beetles in the frames, or webbing from wax moths.

3.2 Seasonal Tool Usage

Beekeeping requires adaptability, as each season brings unique challenges and demands specific tools and strategies to support the hive's needs. The tools a beekeeper relies on in spring and summer may differ significantly from those used in fall and winter. This section covers essential tool usage and seasonal practices to maintain colony health year-round.

Spring and Summer: Frames, Hive Tool, Smoker, and Queen Excluder

The active beekeeping season begins in spring and carries through to late summer, as temperatures warm and floral sources become abundant. During this period, colonies focus on brood rearing and honey production, so beekeepers should conduct frequent inspections and make adjustments as needed.

- **Frames:** With the onset of spring, beekeepers should inspect frames for remaining honey stores, brood patterns, and signs of early nectar flows. New or replacement frames may be added as the colony expands, ensuring there is ample space for brood and honey storage.

- **Swarm Prevention:** As the colony grows, beekeepers may need to add additional brood boxes or supers to prevent overcrowding, a common trigger for swarming behavior. Expanding the hive by adding more frames and boxes keeps the colony busy, reducing the likelihood of swarming.

- **Hive Tool and Smoker:** Both of these tools see frequent use during spring and summer inspections. The hive tool's prying and scraping functions are essential for separating boxes and removing any accumulated propolis or excess wax. The smoker remains a vital tool for calming bees as inspections become more frequent.

- **Queen Excluder:** As honey production ramps up, beekeepers typically add a queen excluder between the brood box and honey supers. The excluder prevents the queen from laying eggs in the honey supers, ensuring that honey stores remain brood-free and ready for extraction.

- **Fall and Winter:** Hive Insulation, Varroa Mite Treatments, and Mouse Guards
  As fall approaches, beekeepers shift their focus to preparing the hive for winter. Insulating the hive, controlling pests, and securing the structure against potential predators all become priorities to protect the colony from harsh winter conditions.

- **Hive Insulation:** In colder climates, beekeepers often insulate their hives to help bees maintain a stable internal temperature. Insulation techniques may include wrapping the hive in breathable insulation materials, adding moisture control boards, and using inner covers to retain warmth. A well-insulated hive reduces stress on the bees and conserves their energy as they form a cluster to stay warm.

- **Varroa Mite Treatments:** Varroa mites pose a severe threat to bee health, especially as bees enter the winter season with weaker immune systems. Mite treatments, such as oxalic acid vaporization or thymol-based solutions, are commonly applied in fall to reduce mite populations before winter. Effective varroa control can significantly improve a colony's survival rate, as high mite loads can lead to weakened bees and increased susceptibility to disease.

- **Mouse Guards:** During winter, rodents like mice may seek shelter inside hives, posing a risk to both hive structure and colony health. Installing a mouse guard at the entrance helps prevent mice from entering the hive. Mouse guards are typically metal or plastic screens that allow bees to pass freely while blocking larger pests. They are simple to install and provide vital protection during colder months when hive entrances are less actively patrolled by bees.

By following structured inspection routines and adapting tool usage to seasonal needs, beekeepers can ensure their colonies remain healthy, resilient, and productive year-round. The hive tool, smoker, bee brush, queen excluder, and seasonal equipment like hive insulation and mouse guards all contribute to effective hive management, promoting colony stability and growth.

Chapter 4: Protecting Bee Health and Hive Safety

In beekeeping, hive protection and bee health are essential for a thriving colony. Various threats, natural and environmental, challenge the safety and resilience of hives, and they demand strategic responses. Effective protection begins with understanding the range of potential risks, from small pests to large predators and fluctuating environmental conditions. By implementing chemical-free strategies and prioritizing hive hygiene, beekeepers can maintain a balanced, healthy hive ecosystem that fosters bee productivity and longevity.

4.1 Natural Hive Threats

In the natural world, bee colonies face several threats from pests and predators. Some pests target the internal structure and resources of the hive, while larger predators may seek out hives for honey or as a food source. Beekeepers can adopt a range of strategies to protect their colonies from these threats, using preventative and reactive measures to maintain a safe hive environment.

Pests: Wax Moths, Varroa Mites, and Small Hive Beetles

Wax Moths: Wax moths are notorious for infiltrating weak or unattended hives, laying eggs within the wax comb. Once the eggs hatch, wax moth larvae feed on the wax, honey, and pollen, and they tunnel through the comb, which disrupts the hive structure.

- **Preventative Measures:** Regular hive inspections help detect signs of wax moth infestation early. Keeping hives strong and healthy reduces the risk, as wax moths are more likely to invade weak colonies. Additionally, using mechanical controls like traps can help capture moths before they enter the hive.

- **Managing Infestations:** In cases of wax moth infestation, affected frames can be removed and frozen to kill larvae, while damaged frames should be cleaned or replaced to restore hive structure.

Varroa Mites: Varroa mites are one of the most harmful pests to honey bees, as they attach themselves to adult bees and developing brood, feeding on their bodily fluids and weakening them. A heavy varroa infestation can lead to significant colony losses if left unchecked.

- **Monitoring:** Regular mite counts help beekeepers assess the mite load within a colony. This can be done through techniques like sugar dusting or alcohol washes, which allow beekeepers to estimate the infestation level.

- **Natural Treatment Methods:** Varroa mites are often managed with chemical-free treatments, such as dusting with powdered sugar to dislodge mites or using essential oils that deter mites without harming the bees. Additionally, drone comb trapping can help remove mites from the hive.

Small Hive Beetles (SHB): The small hive beetle, a pest native to Africa, can cause severe damage in areas where colonies are unaccustomed to its presence. SHB adults and larvae feed on hive resources, creating an acidic environment that contaminates honey stores.

- **Prevention and Control:** Beetle traps and beetle boards are commonly used to capture adult beetles. Keeping colonies strong and limiting hive space helps reduce the chance of SHB infestation, as strong colonies are better able to defend against intruders.

Predators: Raccoons, Skunks, and Bears

In addition to insect pests, beekeepers must sometimes contend with larger predators who may view hives as a source of food. While these predators usually target hive resources rather than bees directly, their attacks can disrupt and damage the hive, forcing the colony to expend valuable energy on repair and recovery.

Raccoons and Skunks: Smaller predators like raccoons and skunks may scratch at hives to eat bees or scavenge honey. They are nocturnal and typically approach hives at night when bees are less active.

- **Preventative Strategies:** Raising the hive on a sturdy stand deters skunks, as they tend to feed on ground level. For raccoons, using secure hive latches or placing a wire mesh barrier around the hive can help prevent these animals from accessing the hive.

Bears: In regions where bears are present, they pose a major threat to apiaries, as they can overturn and destroy hives to access honey and brood.

- **Bear Fencing:** Electric fencing is one of the most effective deterrents against bears. Positioning a fence around the hive area with electrified wires spaced close together ensures that bears are discouraged without causing harm.

- **Hive Positioning:** Keeping hives in open areas away from dense vegetation also helps, as bears are less likely to approach hives in spaces with limited cover.

4.2 Environmental Threats

Beyond direct threats from pests and predators, environmental factors also influence hive health and safety. Temperature fluctuations, precipitation, and seasonal changes can all place stress on a hive. By preparing for these conditions, beekeepers can help maintain a stable environment within the hive that supports the bees' natural resilience.

- **Temperature Extremes:** Insulating the Hive and Providing Ventilation

- **Managing Heat in Summer:** During warmer months, excessive heat can put stress on the colony, causing bees to spend extra energy cooling the hive. Adding ventilation, like screened bottom boards or placing small spacers at the top of the hive, improves airflow and reduces overheating.

- **Shade and Water Sources:** Positioning hives in partial shade and providing a nearby water source, such as a shallow dish with rocks, can further support bees' cooling efforts.

- **Insulation in Winter:** Insulating hives for winter is critical in colder climates. A well-insulated hive helps bees maintain a consistent temperature, reducing the energy they expend to keep the brood area warm.

- **Insulation Materials:** Wrapping hives in breathable, insulating materials, such as tar paper or foam boards, helps conserve warmth. Moisture control boards or absorbent inner covers are also recommended to manage condensation inside the hive, which can chill bees.

- **Flood and Rain Protection:** Hive Stand Placement and Rain Covers
Heavy rains and floods present serious risks to hives, as excess moisture can cause mold growth, weaken hive structure, and lead to brood chill.

- **Elevating Hives:** Positioning hives on sturdy, elevated stands keep them off the ground and away from pooling water. This setup helps prevent water from seeping into the hive and maintains a dry, stable foundation.

- **Rain Covers and Roofing:** Adding a small roof or rain cover can prevent excess water from dripping into the hive. Beekeepers can also use slanted hive covers to direct water away from the hive entrance, reducing the chance of moisture buildup.

4.3 Chemical-Free Hive Health Strategies

A chemical-free approach to hive health aligns with the bees' natural environment and reduces the risk of chemical residues affecting honey quality. Strategies like Integrated Pest Management (IPM) and consistent beekeeping hygiene offer effective alternatives to synthetic treatments, helping beekeepers maintain a balanced hive ecosystem.

- **Integrated Pest Management (IPM):** Reducing Pests Without Chemicals IPM is a holistic approach that combines physical, mechanical, and biological controls to manage pest populations without relying on chemicals.

- **Monitoring and Early Detection:** IPM emphasizes regular pest monitoring to detect infestations early, which enables beekeepers to take targeted actions. Monitoring tools, such as sticky boards or mite counts, provide insight into pest levels and help guide treatment decisions.

- **Mechanical Controls:** Mechanical methods, such as screened bottom boards and drone comb trapping, are central to IPM practices. Screened bottom boards allow mites to fall out of the hive, reducing mite levels naturally. Drone comb trapping, where drone comb is removed and frozen, helps control varroa mite populations, as mites preferentially lay eggs in drone cells.

- **Biological Controls:** Certain biocontrol agents, like beneficial fungi or essential oils, are sometimes introduced to manage pests. Essential oils like thymol and tea tree oil can deter pests and even support bees' immune systems. However, these should be used with caution, as high concentrations may harm bees if misapplied.

- **Beekeeping Hygiene:** Regular Cleaning, Tool Disinfection, and Natural Mite Treatments Maintaining high standards of hygiene in the apiary minimizes the risk of disease transmission and fosters a healthier hive environment.

- **Cleaning Hive Components:** Regular cleaning of hive components, such as the frames, bottom boards, and entrances, prevents debris buildup and reduces the chance of disease. Scraping off old wax and propolis from frames or using a mild vinegar solution to clean bottom boards helps maintain hive cleanliness without introducing chemicals.

- **Tool Disinfection:** Tools used in the hive, like the hive tool, frame gripper, and uncapping knife, should be disinfected frequently. Boiling tools in hot water or using a mild bleach solution effectively removes pathogens. Avoiding shared tools between hives, or disinfecting them before each use, can prevent disease spread within the apiary.

- **Natural Mite Treatments:** Beekeepers can turn to organic acids, like oxalic and formic acid, for mite control. While naturally occurring, these acids require careful application and correct timing to avoid harming the colony. Dusting bees with powdered sugar is another chemical-free option, as it encourages grooming behavior that can reduce mite levels.

By implementing protective measures against natural and environmental threats and adhering to chemical-free health practices, beekeepers support the colony's resilience. Through an integrated approach combining prevention, maintenance, and hygiene, beekeepers can maintain a balanced, healthy hive environment where bees are empowered to thrive.

Chapter 5: Hive Monitoring and Technology Tools

In recent years, beekeeping has embraced technological advancements that allow for a more thorough, less invasive way to monitor and support hive health. Modern monitoring tools, from digital hive scales to infrared imaging, help beekeepers collect valuable insights into hive activity and colony well-being without disturbing the bees. This chapter explores the main tools and technologies that beekeepers can use to improve hive monitoring practices, enhance productivity, and proactively respond to potential threats.

5.1 Digital Hive Scales

Digital hive scales provide beekeepers with real-time insights into colony productivity and hive health. By tracking fluctuations in hive weight, beekeepers can monitor nectar flow, honey production, and colony growth across the seasons. The ability to analyze weight trends offers a powerful tool to gauge hive health and respond to environmental changes impacting nectar availability.

Monitoring Weight: Real-Time Tracking of Nectar and Honey Production

- **Tracking Nectar Flow:** Seasonal nectar flow is critical for honey production, and monitoring hive weight reflects changes in nectar availability. During high nectar flow, the hive weight will increase rapidly as bees collect and store nectar, while a decline in weight may indicate reduced nectar availability or an increase in foraging during food shortages.

- **Assessing Honey Reserves:** Monitoring hive weight allows beekeepers to assess honey reserves without frequent inspections, which can disturb the bees. A sudden drop in weight, for example, could indicate the colony is consuming stored honey, signaling to beekeepers that supplemental feeding may be necessary.

Seasonal Analysis: Analyzing Colony Productivity Trends

- **Annual Hive Performance:** Tracking hive weight data over months or years provides beekeepers with a comprehensive overview of colony productivity across seasons. This information can reveal trends, such as when nectar flows typically begin and end, allowing for better planning.

- **Assessing Hive Location and Environment:** Weight trends can also help beekeepers evaluate the suitability of a hive's location. For example, if hive weights are consistently low, it may indicate a poor forage area, prompting the beekeeper to consider relocating the hive for better access to nectar sources.

5.2 Temperature and Humidity Sensors

Temperature and humidity are essential indicators of hive health. Monitoring these internal conditions can provide invaluable insights into colony status, as variations in temperature and humidity often correlate with brood health, colony stress, and bee activity. Digital temperature and humidity sensors enable remote monitoring, allowing beekeepers to address issues without disturbing the hive.

Internal Conditions: Monitoring Hive Conditions Remotely

- **Temperature:** A stable temperature within the brood chamber is essential for healthy brood development. By using sensors to monitor temperature, beekeepers can identify deviations that may indicate potential issues, such as poor insulation, insufficient hive ventilation, or colony stress.

- **Humidity:** Maintaining proper humidity levels is also crucial, as excessive moisture can lead to mold growth, while too little humidity can stress the colony. Ideal humidity levels help bees maintain wax comb integrity and manage brood development. Humidity sensors allow beekeepers to detect imbalances and adjust hive conditions as needed.

Health Indicators: Using Data to Detect Colony Distress

- **Detecting Queen Issues:** Changes in temperature can sometimes indicate the loss of a queen. If the queen dies or is failing, bees may reduce brood production, which can lower the temperature of the brood area. Monitoring brood temperature trends can alert beekeepers to investigate potential queen issues.

- **Colony Stress Signals:** Variations in temperature and humidity patterns can signal colony stress caused by external factors, such as a nearby pesticide application or predator disturbance. By observing these changes, beekeepers can take preventive measures, such as providing additional ventilation, installing hive covers, or relocating hives temporarily.

5.3 Infrared Imaging

Infrared imaging, also known as thermal imaging, has become a valuable tool for beekeepers. This technology allows for non-invasive hive monitoring by creating thermal maps that reveal colony density, brood health, and hive invaders. By analyzing heat signatures, beekeepers can identify potential issues without opening the hive, which minimizes disturbances and helps maintain hive stability.

Thermal Maps: Identifying Weak Colonies and Brood Health Without Opening the Hive

- **Assessing Colony Strength:** A strong colony typically generates a more concentrated heat signature in the brood area due to the bees clustering around brood to maintain temperature. Conversely, a weak colony may produce a dispersed heat pattern, signaling low bee density or insufficient heat production. Beekeepers can use these thermal maps to identify colonies that need intervention, such as feeding or insulation, to strengthen their numbers.

- **Brood Health Monitoring:** Infrared imaging can also help assess brood health, as healthy brood clusters are warmer than peripheral areas of the hive. If thermal images show unusual heat patterns, it may indicate issues like a brood disease or poor brood density, prompting a more detailed inspection to diagnose the cause.

Detecting Robbing Behavior: Recognizing Hive Invaders Through Heat Signatures

- **Identifying Robbing Activity:** During robbing incidents, bees from other colonies enter the hive to steal honey, creating heat patterns near the hive entrance and potentially disturbing the temperature inside. By examining the heat signatures, beekeepers can detect unusual activity at the hive entrance and take action, such as reducing the entrance size or moving the hive to discourage robbers.

- **Detecting Predators or Pests:** Thermal imaging can reveal other types of invaders, such as mice or small mammals that might enter the hive seeking warmth or food. This is particularly valuable in colder months when small animals might seek refuge in hives. By spotting these intrusions early, beekeepers can take steps to reinforce the hive and deter predators.

Leveraging Technology for Hive Monitoring

The rise of digital and thermal tools has transformed hive management, allowing beekeepers to monitor their hives remotely and in real-time. By integrating these technologies, beekeepers gain critical insights that enable them to respond proactively to threats, improve hive productivity, and maintain bee health with minimal disturbance.

- **Data-Driven Decisions:** The data gathered from hive scales, sensors, and thermal imaging empower beekeepers to make informed decisions. For instance, if weight data shows declining honey reserves, they can plan supplemental feeding, while temperature data can inform hive placement to optimize seasonal health.

- **Reduced Hive Disturbances:** Traditional hive inspections often involve opening hives, which disrupts bees and exposes them to environmental stressors. Digital tools allow beekeepers to track hive conditions without intrusion, supporting a more natural hive environment and reducing stress on the colony.

- **Future Innovations:** As technology advances, new tools are emerging that integrate artificial intelligence (AI) and machine learning to analyze hive data. These systems can detect patterns that humans may overlook, offering predictive insights into colony health, forage availability, and seasonal trends. Looking ahead, such technologies hold the potential to further enhance sustainable beekeeping practices and support bee conservation on a global scale.

Through the careful use of digital monitoring tools and data-driven practices, beekeepers can foster healthier, more productive colonies. In the face of environmental challenges, the combination of technology and traditional knowledge equips beekeepers with the skills and information needed to safeguard the future of honeybees and promote a thriving ecosystem.

Chapter 6: Hive Maintenance and Re-Queening

Proper hive maintenance and management are essential to ensure that colonies remain healthy, productive, and resilient. As frames wear down and queens age, beekeepers must take proactive steps to replace frames and introduce new queens to maintain colony vigor. This chapter delves into the detailed practices of frame and foundation management, the art of re-queening, and how each approach contributes to a thriving bee colony.

6.1 Frame and Foundation Management

Frames and foundations are the building blocks of a hive, providing bees with the structure to build their comb and store food, raise brood, and carry out essential colony functions. Over time, however, frames and foundations degrade or become overused, requiring beekeepers to intervene to maintain a functional and healthy hive.

Replacing Worn Frames: Extending Hive Life by Regular Frame Replacement

Signs of Wear and Tear: Frames can accumulate a significant amount of propolis, debris, and old comb over time. Worn frames are darker in color, often brittle, and may harbor pathogens or pests. Regular inspection of frames allows beekeepers to assess when frames are past their prime and need replacement.

- **Benefits of Fresh Frames:** Replacing old frames with fresh ones reduces the colony's exposure to pests, pathogens, and contaminants. Fresh frames offer bees a cleaner, more sanitary environment, essential for healthy brood development. New frames also encourage more efficient comb building, as bees can lay out new cells in a way that optimizes brood and honey production.

- **Rotation Schedule:** Many beekeepers adopt a rotational schedule, replacing one or two frames each year in each box. This gradual approach prevents major disruptions and allows bees to adjust naturally to the new frames. By implementing a regular rotation, beekeepers maintain hive hygiene and structure without overburdening the bees with too many changes at once.

Wax Foundation: Installing Fresh Foundation for Comb Building

- **Foundation Basics:** The foundation provides bees with a template on which to build comb, guiding the hexagonal cell formation and enhancing the hive's overall structure. Wax foundation is the most common choice, as it mimics natural bee environments and is readily accepted by bees.

- **When to Replace Foundation:** Over time, foundation wax can become worn or brittle, and cells may become irregular. When foundation shows signs of damage, beekeepers should replace it to encourage healthy and efficient comb building. Fresh foundation also reduces the risk of pest buildup, as comb degradation can invite wax moths and other pests.

- **Types of Foundations:** Some beekeepers prefer plastic foundation, which lasts longer and provides added durability. However, wax foundation is often favored for its natural properties and acceptance by bees. Beekeepers should choose the foundation type based on their hive management preferences and goals.

6.2 Re-Queening

A strong, healthy queen is the heart of a thriving colony. Over time, however, queens can decline in productivity, resulting in reduced brood production, poor colony health, and increased susceptibility to disease. Re-queening, or introducing a new queen, is a vital management practice for maintaining colony strength and stability.

Identifying a Failing Queen: Recognizing Signs of Queen Decline

- **Reduced Brood Patterns:** One of the first signs of a failing queen is a noticeable decline in the brood pattern. A healthy queen lays eggs in a dense, uniform pattern, while a failing queen will leave patchy areas, leading to gaps in the brood comb. Patchy brood patterns indicate that the queen's egg-laying ability is declining, potentially affecting the colony's population growth and productivity.

- **Drone Brood Production:** If a queen is nearing the end of her reproductive lifespan, she may lay more drone (male) eggs instead of worker (female) eggs. An abundance of drone brood can signal that the queen is no longer producing fertile eggs, which weakens the colony, as drones do not contribute to foraging or brood care.

- **Colony Behavior Changes:** The bees themselves often sense when a queen is failing. They may become agitated, produce fewer honey stores, or begin building "queen cups" or "supersedure cells," where they plan to rear a new queen. Observing these behaviors can help beekeepers identify when it's time to re-queen the hive.

Successful Re-Queening: Techniques for Introducing a New Queen

- **Timing the Re-Queening Process:** The best time to re-queen a hive is during spring or early summer when resources are plentiful, and the colony is at its strongest. However, a failing queen should be replaced as soon as possible, even if it means re-queening outside the ideal season. For best results, aim to re-queen during a period of favorable weather to give the new queen time to be accepted by the colony.

- **Preparing for Re-Queening:** Begin by removing the old queen. A hive cannot support two queens, so the existing queen must be removed before introducing a new one. It's important to ensure the colony has not begun creating supersedure cells;

otherwise, they may reject the new queen in favor of a queen they have raised themselves.

- **Using a Queen Cage:** When introducing a new queen, place her in a queen cage with a candy plug that worker bees will slowly chew through, releasing the queen into the hive over a few days. This gradual release process allows the colony time to become accustomed to her pheromones, increasing the likelihood of acceptance. The cage should be positioned between frames in the brood chamber for optimal introduction.

Observing Colony Behavior: Once the queen is introduced, monitor the colony's behavior closely. If the workers appear aggressive toward the cage, it may indicate resistance to the new queen. Give the hive additional time, but if aggression continues, you may need to re-evaluate the re-queening process.

The Role of Maintenance and Re-Queening in Long-Term Hive Health

Regular hive maintenance and strategic re-queening are essential to the health and productivity of a colony. Frames and foundations provide the structure for a thriving colony, while a healthy queen is the key to robust brood production and colony stability.

- **Preventing Hive Decline:** Proactive frame management reduces the accumulation of pathogens and provides bees with a safe, clean environment. Coupled with re-queening practices, these strategies prevent hive decline and support the colony's long-term health.

- **Improving Colony Productivity:** A productive queen ensures a steady supply of worker bees, which in turn boosts the colony's foraging efficiency and honey production. Regular re-queening prevents the drop in productivity that often accompanies a queen's natural aging process.

- **Encouraging Resilience:** Well-maintained hives and strong queens contribute to the resilience of a colony. As environmental stressors continue to affect bee populations, diligent hive management helps beekeepers support colonies through challenging conditions, from pest infestations to extreme weather.

- **Creating a Healthy Environment:** A properly managed hive reduces the need for chemical treatments and interventions, allowing beekeepers to embrace a more sustainable approach to beekeeping. Clean frames, fresh foundation, and a productive queen together foster a balanced and healthy colony ecosystem.

Through regular hive maintenance and timely re-queening, beekeepers can effectively support their colonies' growth, resilience, and productivity. These practices, rooted in beekeeping tradition, continue to be essential in modern beekeeping, combining the wisdom of past generations with the challenges of the present.

Chapter 7: Hive Security and Protecting Against Hive Theft

As beekeeping grows in popularity and demand for honey and pollination services rises, beekeepers face an increasingly common threat: hive theft. Hive security has become a priority, particularly for commercial beekeepers who manage large apiaries and transport hives to different locations for pollination contracts. This chapter provides a comprehensive look at measures beekeepers can take to protect their hives from theft, legal recourse, and insurance options that can mitigate financial loss if theft occurs.

7.1 Preventing Hive Theft

Preventing hive theft requires a combination of strategic placement, physical barriers, and technology. Each measure, when carefully implemented, reduces the likelihood of theft and provides beekeepers with greater peace of mind.

- **Location Tips:** Placing Hives Out of Sight and Setting Up Barriers
  Choosing a Discreet Location: Keeping hives out of public view is the first line of defense against theft. When hives are placed near main roads, high-traffic areas, or popular trails, they become more vulnerable to opportunistic thieves. Selecting a hidden or remote location within a property, such as behind natural barriers like trees, hills, or bushes, can help to conceal the hives.

- **Fencing and Barriers:** Physical barriers are essential, especially for apiaries located on open farmland or near accessible pathways. Fences deter both human and animal intruders. Tall fencing and secured gates make it harder for thieves to quickly access hives, providing an additional layer of security. Electric fencing can be particularly effective, as it deters larger predators like bears while also discouraging trespassing.

- **Controlled Access:** For commercial beekeepers, limiting access to hive locations is crucial. Installing gates that require key codes or locks can prevent unauthorized access, especially when apiaries are located on leased land or open fields. Signs indicating restricted access or surveillance can also be effective deterrents.

Identification: Marking Hives with Unique IDs and Installing Tracking Devices

- **Hive Marking:** Identifying hives with unique markings, such as branded or painted initials, numbers, or logos, serves two purposes. First, it makes it more difficult for thieves to sell stolen hives, as the marked hives are recognizable. Second, it aids in the recovery process, allowing law enforcement to confirm ownership if stolen hives are located. Using waterproof paint or branded initials ensures that markings are visible and durable.

- **Hive Tracking Technology:** Modern tracking devices, such as GPS trackers, are valuable tools in hive theft prevention. Small GPS units can be concealed within hive boxes or attached to bottom boards. Once activated, these devices provide real-time location data, allowing beekeepers to track stolen hives and assist law enforcement in recovery. Many GPS tracking devices come with long battery life and can be remotely monitored via smartphone apps, offering beekeepers a practical solution to secure their investment.

7.2 Legal and Insurance Options

Although prevention is the best defense against hive theft, insurance and legal measures provide an essential safety net. Beekeepers can explore insurance options, keep clear documentation, and use legal recourse if theft does occur.

Bee Hive Insurance: Protecting Your Investment

- **Understanding Hive Insurance:** Beekeeping insurance, particularly hive-specific policies, is increasingly available to protect beekeepers' investments in case of theft, vandalism, or natural disasters. Hive insurance often covers the cost of hive replacement, bee colonies, and equipment. Some policies may also include coverage for loss of honey or wax production and liability protection if the hives are on public or leased land.

- **Choosing the Right Coverage:** Beekeepers should select insurance coverage based on their scale of operation and specific needs. For small-scale beekeepers with only a few hives, general farm insurance policies may be sufficient. However, commercial beekeepers with large apiaries or those involved in pollination contracts should consider comprehensive coverage that includes theft and vandalism. Understanding the policy's terms, exclusions, and limits is essential to ensure adequate protection.

- **Documenting Hive Ownership:** Keeping detailed records of hive locations, equipment inventory, and any unique hive markings is beneficial in the event of a theft. Taking photographs of hives, including close-ups of markings, serial numbers, or branded initials, provides valuable evidence that can support an insurance claim or assist law enforcement. Updated records of hive locations, especially when moved for pollination contracts, streamline the documentation process.

Tracking Technology: Using GPS Tracking to Recover Stolen Hives

- **GPS as a Theft Deterrent and Recovery Tool:** GPS technology has become an effective solution for recovering stolen hives. Many commercial GPS tracking devices are designed specifically for hive monitoring, with features such as motion detection alerts and real-time location tracking. If a hive is moved without authorization, the beekeeper receives a notification, enabling prompt action and increasing the chances of recovery.

- **Selecting a Suitable GPS Device:** There are various GPS trackers on the market, each with different features and price points. Beekeepers should choose devices with long battery life, waterproof casings, and real-time tracking capabilities. Some devices also include geofencing options, where beekeepers can set a boundary around the hive location and receive alerts if the hive crosses that boundary.

- **Working with Law Enforcement:** In the unfortunate event of a theft, GPS data can be invaluable to law enforcement. Beekeepers should be prepared to provide tracking data and any recorded location history to aid in recovery efforts. Additionally, contacting local authorities about the theft immediately can enhance the likelihood of a successful recovery, as early notification is often key.

The Importance of Hive Security and Theft Prevention

Hive theft can be devastating, not only financially but also in terms of the time, care, and commitment beekeepers invest in their colonies. Implementing a strong security plan, which includes a mix of preventive measures, identification, and technology, minimizes the risk of theft and helps beekeepers protect their investments.

- **Cost-Effective Security:** While hive tracking technology and fencing may seem like additional expenses, they often pay off by reducing the risk of significant financial loss due to theft. Preventive measures are especially valuable for commercial beekeepers who have invested heavily in their hives.

- **Contributing to Hive Recovery:** By marking hives and using tracking technology, beekeepers make it easier for law enforcement to identify stolen property, helping deter future thefts. Visible identification and the possibility of GPS tracking act as deterrents, discouraging thieves who might otherwise see hives as easy targets.

- **Raising Awareness and Community Involvement:** Hive theft prevention can also benefit from community involvement. Establishing relationships with nearby landowners, neighbors, and other beekeepers fosters a supportive network that can help monitor hive locations. Community awareness of hive theft helps discourage criminal activity and increases the likelihood of identifying suspicious behavior around hive sites.

Hive Theft and the Future of Beekeeping Security

As beekeeping continues to grow, the threat of hive theft remains a challenge for the industry. However, advancements in technology, coupled with accessible insurance options and preventive practices, equip beekeepers with the tools they need to secure their hives.

Through diligent planning, regular updates to security measures, and community engagement, beekeepers can effectively protect their hives and ensure the safety of their colonies. With awareness and the right strategies, beekeeping can remain a rewarding and sustainable practice, even in the face of modern challenges like hive theft.

Chapter 8: Sustainable Beekeeping and Community Impact

In recent years, sustainable beekeeping has become a focal point within the beekeeping community. As environmental awareness grows, more beekeepers are recognizing the importance of eco-friendly practices that protect bee health and promote a balanced ecosystem. Beyond the personal or commercial rewards of beekeeping, many are taking their efforts a step further by engaging with their communities to educate the public on the critical role of pollinators. In this chapter, we'll explore sustainable beekeeping practices, the impact of these practices on bee health, and the role of community engagement in fostering a deeper understanding of beekeeping's benefits.

8.1 Eco-Friendly Beekeeping Practices

Sustainable beekeeping goes beyond simply managing hives. It involves mindful choices that minimize environmental impact, protect bee health, and create a supportive environment for pollinators to thrive. For the beekeeper, eco-friendly practices may also lead to healthier colonies, more resilient to disease and pests, and capable of producing higher-quality honey.

Minimizing Chemicals: Organic Pest Control Methods

Organic Varroa Mite Control: Varroa mites are one of the most significant threats to bee colonies worldwide. Instead of relying on harsh chemicals, sustainable beekeepers use organic treatments such as oxalic acid vapor, formic acid, and essential oils, which can help control mite populations without harming bees. Regular monitoring through mite count tests enables beekeepers to catch infestations early and intervene only as needed, further reducing the need for chemical treatments.

- **Integrated Pest Management (IPM):** IPM is an environmentally friendly approach to pest control that combines biological, physical, and cultural methods. For example, beekeepers can install screened bottom boards, which allow mites to fall out of the hive rather than remain inside, naturally reducing their numbers. IPM promotes a holistic strategy, balancing pest control without heavy reliance on synthetic chemicals.

- **Use of Natural Comb Foundation:** Some beekeepers avoid using synthetic wax foundation and opt for natural or foundationless frames instead. This allows bees to create their comb structures, minimizing the risk of chemical contaminants found in commercial wax. By avoiding wax treated with pesticides or antibiotics, beekeepers reduce the likelihood of exposing bees to potentially harmful residues.

Bee Health: Avoiding Stress Through Sustainable Practices

- **Location and Hive Placement:** Sustainable beekeeping includes careful consideration of hive location. Placing hives in diverse environments with access to various pollen and nectar sources promotes colony health and reduces stress caused by limited resources. Beekeepers can also rotate hives to different locations as needed, especially in areas where crops may be treated with pesticides, to protect bees from chemical exposure.

- **Minimal Hive Interventions:** Beekeepers practicing sustainable methods understand that frequent hive inspections and relocations can disrupt colony harmony. By only intervening when necessary, beekeepers reduce stress on the bees and allow them to carry out their natural behaviors, fostering a more resilient colony. This low-intervention approach supports hive stability and allows bees to focus on brood-rearing, foraging, and honey production.

- **Natural Overwintering Techniques:** Sustainable beekeepers prioritize methods that support bees' natural wintering instincts. Instead of artificial feeding or heating, which can stress the colony, they ensure adequate honey reserves before winter and insulate hives to preserve warmth. This approach encourages bees to use their own stores of honey and propolis to sustain themselves through winter, mimicking their natural behavior.

8.2 Community Engagement and Pollinator Education

Beekeeping can serve as a gateway to broader environmental awareness within communities. Beekeepers are in a unique position to educate others on the essential role of bees in maintaining biodiversity and food security. Through workshops, educational outreach, and community events, beekeepers can inspire others to protect pollinators and consider sustainable practices in their own lives.

Educating on Bees: Community Outreach on the Importance of Bees

- **Raising Awareness:** Bees are critical to our ecosystems, but their importance is often overlooked. Beekeepers can address this by sharing insights into the life of the hive, the pollination process, and the relationship between bees and agriculture. Educational events that explain how honeybees contribute to biodiversity and why pollinators are crucial to local agriculture can shift public perception and highlight the need to protect bees.

- **School Programs:** Many beekeepers partner with schools to introduce children to beekeeping and environmental stewardship. Engaging young learners through interactive activities, such as observing bee behavior or identifying bee-friendly plants, fosters a sense of responsibility towards pollinators. School programs that incorporate bees can also inspire future beekeepers and conservationists, nurturing a long-term appreciation for pollinators.

- **Creating Pollinator-Friendly Spaces:** Beekeepers can guide community efforts to create pollinator-friendly spaces, including planting native flowers, setting up community gardens, and reducing pesticide use in public areas. These initiatives help bees thrive and provide habitats for other pollinators like butterflies and native bees. Public spaces designed with pollinators in mind can serve as a model for private gardens and encourage residents to adopt bee-friendly practices.

Hosting Workshops: Sharing Knowledge on Beekeeping Basics and Bee Protection

- **Beekeeping 101:** Community workshops offer an accessible introduction to beekeeping, making it less intimidating for newcomers. Through these workshops, aspiring beekeepers can learn the basics, including hive setup, bee biology, and responsible hive management. These sessions demystify the process and empower individuals to start their own hives if they wish to do so.

- **Advanced Beekeeping Techniques:** For those with some experience, workshops on specific topics, such as natural mite control, hive design, and honey extraction, allow for deeper engagement. Advanced classes help intermediate beekeepers hone their skills while reinforcing sustainable practices.

- **Workshops on Pollinator-Friendly Gardening:** Many people are interested in supporting bees without becoming beekeepers. Gardening workshops focused on native plants, flower diversity, and pesticide-free techniques can teach community members to create bee-friendly environments. Beekeepers can also explain which plants provide the most nourishment for bees throughout the seasons and why diversity in plant species is essential for healthy pollinator populations.

The Lasting Impact of Sustainable Beekeeping and Community Engagement

Sustainable beekeeping isn't just about protecting bees; it's about cultivating a culture that values biodiversity, responsible stewardship, and community resilience. By adopting eco-friendly beekeeping practices, beekeepers contribute to the health of the environment and set an example for those around them. When combined with community outreach and education, these practices have a ripple effect, inspiring others to support pollinators and consider their own role in conservation.

- **Building Stronger Ecosystems:** Sustainable beekeeping practices support biodiversity, help maintain healthy soil, and encourage the growth of native plants. Healthy bee populations also improve the resilience of other local wildlife and plants, fostering a more balanced ecosystem that benefits the community as a whole.

- **Fostering Community and Connection:** By sharing their knowledge and engaging with others, beekeepers create connections that strengthen community bonds. These efforts remind people of their interdependence with nature and the importance of collaborative conservation. Communities that prioritize pollinator health are not only healthier themselves but are also more invested in preserving their local environment.

- **Creating Environmental Champions**: Education and outreach transform individuals into advocates for the environment. When people learn about the essential role of bees, they are more likely to take action to protect them, whether through reducing pesticide use, planting pollinator gardens, or supporting local beekeepers. Each new environmental champion represents a step towards a more sustainable future.

A Vision for the Future of Beekeeping

The future of beekeeping lies in the commitment to sustainability and community involvement. By embracing eco-friendly practices and sharing the importance of pollinators, beekeepers are paving the way for a future where bees and biodiversity can thrive. Sustainable beekeeping is a powerful testament to the positive impact individuals can have on their environment and community.

Beekeepers who engage with their communities not only contribute to the preservation of bee populations but also inspire others to protect the natural world. As sustainable practices become the standard and community initiatives flourish, we can envision a future where beekeeping continues to be both a vital agricultural practice and a cornerstone of environmental stewardship.

The lasting legacy of these efforts is a world where bees, plants, and people coexist harmoniously, each playing a role in supporting the delicate balance of nature. Sustainable beekeeping is not merely a method; it is a philosophy rooted in respect, responsibility, and a shared commitment to nurturing our planet. Through these combined efforts, beekeepers and their communities create a foundation for a thriving, resilient future.

Sadiq Ali Al-Qatari: Born on January 1, 1961, in the vibrant city of Qatif, Eastern Saudi Arabia, Sadiq Ali Al-Qatari has woven a rich tapestry of experiences that inform his writing and activism. Raised in Qatif, he completed his primary, intermediate, and secondary education before embarking on a transformative journey to the United States, where he earned a Bachelor of Science in Petroleum Engineering from the prestigious University of Southern California, thanks to the support of Saudi Aramco.

Sadiq enjoyed a successful career as a consultant at Saudi Aramco, where he honed his analytical skills and deepened his understanding of the industry. Upon retiring, he embraced his true passions: writing, research, photography, and social activism. Through his words, he aims to engage and inspire readers, capturing the beauty of the world through his lenses while advocating for positive social change.

With a unique blend of technical expertise and creative vision, Sadiq brings a fresh perspective to his endeavors. His dedication to making a meaningful impact resonates through his work, inviting readers to join him on a journey of exploration and reflection. Whether through the lens of a camera or the pages of a book, Sadiq Ali Al-Qatari is committed to illuminating the world around us.

ISBN 9798345028292

90000

9 798345 028292